The United States National Climate Assessment

NCA Report Series, Volume 1

MIDWEST REGIONAL WORKSHOP

February 22 - 24, 2010
Chicago, Illinois

I0468007

United States Global Change Research Program
National Climate Assessment

NCA Report Series, Volume 1: Midwest Regional Workshop

NCA Report Series

The NCA Report Series summarizes regional, sectoral, and process-related workshops and discussions being held as a part of the Third National Climate Assessment (NCA) process.

The first regional and strategic guidance workshops to contribute to the 2013 NCA were held in Chicago in February 2010. Volumes 1 and 2 of the NCA Report Series summarize the discussions and outcomes of these workshops. A list of planned and completed reports in the NCA Report Series can be found online at http://globalchange.gov/what-we-do/assessment.

CONTENTS

CONTENTS

Overview of the Workshop

The purpose of this meeting was to present stakeholders in the Midwest region with an opportunity to provide input to the development of the strategic plan for the next National Climate Assessment (NCA). The meeting was organized around discussions of (1) up-to-date climate information for the U.S. and for the Midwest, setting the stage for subsequent conversations (2) key sectoral and regional issues and questions, including sources of vulnerability (3) options available to address adaptation and mitigation issues and associated barriers and information needs, and (4) input regarding development of the regional components and stakeholder engagement methods of the NCA. These discussions in turn helped to inform the subsequent strategic planning meeting on the design of the NCA (NCA Report Series, Volume 2).

Many of the participants in this meeting currently are or had been involved in assessments and climate-related decision processes, or had evaluated or managed similar processes. They represented a wide range of sectors, regions, government agencies and universities.

The format of this workshop included both plenary sessions and facilitated breakout sessions. All sessions were recorded and highlights reported back for plenary wrap-up sessions. The agenda is attached as Appendix A and the Participant List is Appendix B.

Taking Action: A Prelude

The First National Assessment: Lessons Learned and Expectations for a New Round of Assessments

One of the great challenges of strategies for delivering climate information is matching user expectations with the capacity to deliver that information. When work began on the First National Assessment in the late 1990s, a partnership of federal agencies conducted regional workshops across the country to engage scientists and stakeholders in identifying the types of climate information needed for making decisions and the best processes and practices for delivering that information. Such inputs continue to drive the National Assessment process and shaped four main goals for subsequent work:

- Establish a set of operational regional networks, using the unique local capacities available, to develop appropriate responses to climate change.
- Foster partnerships involving leaders from federal, state, and local government, business, academia, non-governmental organizations and the general citizenry. It is vitally important that business leaders are involved in this process, and that they view climate change as not only a challenge but also as an opportunity to make money and create jobs. These kinds of partnerships among stakeholders are evolving, as well as the mindset that we should think about bolstering a sustainable environment in partnership with fostering a robust economy.
- Regional networks should work toward solutions to climate change challenges in the context of other pressing issues in their regions. People are interested in solutions, both in terms of mitigation and adaptation. It is important that those people involved in discussions about solutions think about the interactions among mitigation and adaptation options (*e.g.*, could a proposed solution such as biofuels have unintended consequences?). Solutions need to be considered in the context of many issues beyond climate change.
- Design and build an appropriate decision-support process and next generation regional engagement process that is both responsive and flexible. In order to implement solutions, the questions of decision makers must be answered appropriately during the decision making process.

Chicago Climate Action Plan

A growing number of cities, states and countries have established their own planning processes and engaged experts to help them respond to climate change. In the United States, the City of Chicago consulted leading scientists to describe various scenarios for Chicago's climate future and how those would impact life in the City. The resulting report found that impacts of climate change pose great risks to the city's economy and health. A group of leaders in the business, civic, environmental, foundation, and other nonprofit communities worked with local research partners to take on this incredible challenge and turn it into a great opportunity.

The Chicago Climate Task Force agreed that Chicago needs to achieve an 80 percent reduction below its 1990 greenhouse gas emissions level by the year 2050 in order to do its part to avoid the worst global impacts of climate change. To achieve this 80 percent reduction, the task force developed the Chicago Climate Action Plan, which outlines strategies to reduce emissions. The strategies include improving the energy efficiency of buildings, using clean and renewable energy sources, improving transportation options, reducing waste and industrial pollution, and developing adaptation plans to prepare for climate changes that cannot be avoided.

The City's climate adaptation plans are prioritized and implemented based on three criteria. First, the City tries to leverage "business as usual." Since the mid-1980s the city of Chicago has invested in sustainability initiatives in an effort to make it the most environmentally friendly city in the nation and improve the quality of life for city residents. Climate change is one more reason to continue to fund ongoing "green" projects. Second, the City prioritizes

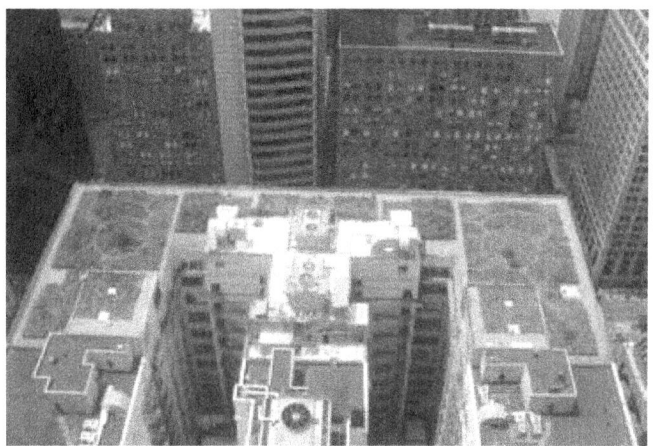

Chicago's City Hall, Green Roof
Photo courtesy of Tony The Tiger and Wikimedia Commons

adaptation projections that have collateral benefits in terms of mitigation. The City's urban heat island reduction projects have both mitigation and adaptation benefits in terms of reducing energy use and greenhouse gas emissions while addressing extreme heat impacts. Finally, the City prioritizes programs and allocates resources according to areas that have greatest needs or are most vulnerable to the impacts of climate change.

The City has already accomplished many of its goals. For example, in the last five years, seven million square feet of green roofing has been installed. The City's extreme weather operations plan now recognizes climate change impacts assessment information, and the urban forestry management plan incorporates urban heat island information. Efforts to monitor alternative roadway materials' performance has influenced the infrastructure materials market and catalyzed business opportunities. The City created a Chicago Climate Action Plan Storm Water Management Framework that establishes effective, natural onsite storm water management tools and a sewer capital investment plan to eliminate water in basements. Finally, the Chicago Trees Initiative is dedicated to increasing Chicago's tree canopy cover to 20 percent by 2020.

Assessing Climate Change in Australia

Australia has seen much success in getting its citizens to recognize that climate change is an issue. Surveys of public opinion show that 84 percent of Australians rank climate change as the greatest threat to Australia's national interest, and 90 percent believe tackling climate change is important. Climate is an important part of Australia's national identity — and it is difficult for citizens to ignore the increasingly severe changes that are taking place.

While Australians are convinced about climate change, the country is still working to effectively answer the question "what do we do about it?" Today, every Australian government department has its own climate change staff, making climate change a part of mainstream policy and development. Climate change is now central to decision making in the government. Despite the country's organizational structure, the amount of money applied toward climate science greatly outweighs the social science, a problem that the United States also faces.

In the past, Australia's national climate assessment approach has been more "climate-centric" than "decision-centric." Assessments identified the

problem but did not provide solutions; too much emphasis was placed on reducing the uncertainty of the science before making decisions. Now, the government is working on approaches to identify the relevant, critical pieces of information that stakeholders need, engaging stakeholders in a discussion of how the government can contribute to their decision process, and then working with them to integrate climate science into their existing decision making systems and strategies.

Australia's national science agency, the Commonwealth Scientific and Industrial Research Organization (CSIRO) established the Climate Adaptation Flagship, a portfolio of research initiatives that bring together researchers, business leaders, government staff, and other stakeholders to link decisions with the best current understanding of climate impacts and adaptation options. During this process, emphasis is placed on finding the opportunities and benefits inherent in adapting to climate change. Australia has a variable climate already; managing the ups and downs of climate is part of business and a sign of excellence.

NOAA Climate Services

In the United States, government agencies are thinking carefully about how they can better respond to the growing demand for more relevant, reliable information about the future state of the climate to allow better planning. The National Oceanic and Atmospheric Administration (NOAA) is working closely with federal, regional, academic and other state and local government and private sector partners to continue to transform science into useable climate services. In February, the agency announced its intent to create a NOAA Climate Service line office. The line office is dedicated to bringing together the agency's strong climate science and service delivery capabilities, and making them more accessible to NOAA partners and other users.

Re-organizing NOAA's climate assets to create the NOAA Climate Service will create a visible and easy to find, single point of entry which will enable more coordinated information sharing. However, no single agency can provide all climate services for all people, and NOAA has only a piece of the information needed. A partnership between federal agencies, local governments, private industry, and all users and stakeholders is needed to provide comprehensive climate research, data collection and dissemination, and climate service provision.

The NOAA Climate Service aims to have a more clearly established regional footprint to coordinate and provide improved regional climate services. Some of the NOAA Climate Service's early priorities include developing a sustained capacity to more effectively provide regional and sectoral climate vulnerability and risk assessments. NOAA is moving quickly to implement the proposed re-organization and hopes to have a functional climate service up and running in FY 2011.

The Impacts of Climate Change in the United States, with a Focus on the Midwest

An Overview of the Changing Climate
In June 2009, the USGCRP released *Global Climate Change Impacts in the United States*.[1] The report summarizes the science and impacts of climate change in the U.S., now and in the future, in accessible, authoritative language. It compiles years of scientific research and accounts for new data not available during the preparation of previous large national and global assessments. The report was produced by a consortium of experts from 13 U.S. government agencies within the USGCRP and from several major universities and research institutes, and went through extensive reviews by the public, a "blue ribbon" panel of experts, and U.S. federal climate agencies.

The report highlights climate change impacts on the U.S. in nine regions and seven sectors. The report concludes with a section titled "An Agenda for Climate Impacts Science," which identifies areas in which scientific uncertainty limits our ability to estimate future climate change and its impacts. The authors of the report received feedback that the content was generally useful, but that, among other suggestions, the next Assessment should focus more on mitigation strategies. The report did not evaluate mitigation technologies or undertake an analysis of the effectiveness of various approaches. The authors believe that the overall communication and outreach strategy for the report needs improvement as well, so that more people are aware of the report, and the capacity to use it is increased.

The following two sections summarize the key overall findings of the report and the findings for the Midwest. Additional information is available in the full report.

Impacts of Climate Change
Key overall findings of the report include the following:
• Global warming is unequivocal and primarily human-induced.
• Climate changes are underway in the United States and are projected to grow.
• Widespread climate-related impacts are occurring now and are expected to increase.
• Climate change will stress water resources.
• Crop and livestock production will be increasingly challenged.
• Coastal areas are at increasing risk from sea-level rise and storm surge.
• Risks to human health will increase.
• Climate change will interact with many social and environmental stresses.
• Thresholds will be crossed, leading to large changes in climate and ecosystems.
• Future climate and its impacts depend on choices made today.

Our Climate Future: The Midwest Region
For the Midwest region, the report finds the following:
• During the summer, public health and quality of life, especially in cities, will be negatively affected by increasing heat waves, reduced air quality, and increases in vector-borne diseases. In the winter, warming will have mixed impacts.
• Significant reductions in Great Lakes water levels, which are projected under higher emissions scenarios, will lead to impacts on shipping, infrastructure, beaches, and ecosystems.
• The likely increase in precipitation in winter and spring, more heavy downpours, and greater evaporation in summer would lead to more periods of both floods and water deficits.
• While the longer growing season provides the potential for increased crop yields, increases in heat waves, floods, droughts, insects, and weeds will present increasing challenges to managing crops, livestock, and forests.
• Native species are very likely to face increasing threats from rapidly changing climate conditions, pests, diseases, and invasive species moving in from warmer regions.

[1] Karl, T.R., J.M. Melillo, T.C. Peterson (eds.). 2009. *Global Climate Change Impacts in the United States*. Cambridge University Press. http://globalchange.gov/what-we-do/assessment/nca-reports.

Critical Sectoral and Regional Issues in the Midwest

Participants identified a number of ecological and socioeconomic sectors that are especially important to consider in the Midwest. Within each sector there are a number of key climate change impacts that the participants highlighted. Participants also identified a number of high-priority, cross-sectoral issues that should be addressed quickly (see Box 1). In general, participants noted that climate change is a risk multiplier – it both exacerbates existing stressors in each sector and brings new impacts which must be dealt with. Climate change is also an environmental justice issue, threatening some populations more than others because of existing advantages or disadvantages. In addition, there are significant interactions among these sectors and with the climate system that must be analyzed and addressed using a system-wide approach.

Aquatic and Wetland Ecosystems

The Midwest spans two of the major watersheds in North America – the Laurentian Great Lakes and the Mississippi River. The Great Lakes and Mississippi River and adjacent coastal areas are important ecologically and economically – for example, as habitat for resident and migrating birds and recreationally- and commercially-important fish species, as a source of drinking water for many municipalities, and as shipping lanes for moving raw materials and finished goods into and out of the Midwest. Changes in temperature and precipitation are likely to have profound impacts on these systems, including changes in water temperature (and thus on the distribution and life histories of warm-, cool-, and cold-water fish species), ice-in and ice-out dates (leading to changes in patterns of coastal erosion due to ice scour or winter storms, the amount of precipitation falling as lake-effect snow *vs.* as rain, increased evaporation during winter months, and safety hazards for ice fishing), and runoff from agricultural and urban areas (resulting in increased levels of sediments and nutrients, which may impact water quality and feed algal blooms that contribute to "dead zones" in the Gulf of Mexico and the Great Lakes). These ecosystems are impacted by a number of stressors, such as invasive species that displace native species and disrupt food webs or changes in land use along the shoreline; taking a multiple stressors approach to identifying vulnerabilities is essential for addressing the impacts of climate change.

Box 1: High-Priority Issues for Adaptation and Mitigation in the Midwest

Issues that participants identified as particularly high-priority for action and investigation included:

• Integrated water management
• Energy as a key driver for decision making
• Finding ways to motivate behavioral change
• Putting a greater emphasis on identifying current impacts of climate change
• Understanding the national security dimensions of climate change
• Promoting ecosystem resiliency
• Identifying the roles for technology in adaptation and mitigation

Terrestrial Ecosystems

Several types of terrestrial ecosystems in the Midwest are vulnerable to climate change, ranging from forests in the north and northeast through savannahs and prairies toward the south and southwest. These ecosystems may play a significant role in both adaptation and mitigation strategies (e.g., as carbon reservoirs and as permanent habitat and migration corridors for wildlife) and important destinations for recreation. These areas, which are already economically important as sources of natural materials (e.g., logging), may become even more important as carbon markets emerge. However, these ecosystems are managed for a wide variety of purposes and by a wide variety of private and public entities. Without a common set of regional goals or best management practices, and without a better understanding of how the ability of these ecosystems to store carbon may change with age and other environmental factors, it will be difficult to determine how best to manage these ecosystems in the face of climate change. Adaptation and mitigation plans must also account for interactions of climate with invasive species and disturbance regimes (e.g., emerald ash borer, wild fires) and should address questions related to both flora and fauna (e.g., impacts of changes in biodiversity on community structure and function, ability of species to migrate northward as temperatures rise, increased potency and growth of nuisance plants such as poison ivy).

Agriculture

In the short-term, many agricultural areas in the Midwest may be perceived as climate change "winners": yields are increasing as warmer spring and fall temperatures extend the growing season (allowing for earlier planting and extended or later harvests) and some crops respond well to higher CO_2

levels. However, other agricultural areas are seeing negative impacts that are only expected to worsen: fruit trees that burst buds as a result of warmer temperatures in early spring are more vulnerable to damage and flower / fruit loss due to late winter and early spring storms; stronger storms and heavier precipitation events throughout the growing season contribute to crop damage and loss. As climate change continues to bring higher temperatures later in the century: livestock may be at risk for heat stress during the summer and will require greater amounts of water; higher volumes of pesticides and herbicides may be required to control insects and weeds (which may also impact water quality and human health nearby and downstream); higher temperatures may reduce crop yields (e.g., seed production in corn); more water may be required for irrigation (affecting both surface water and groundwater supplies). Agricultural practices in the Midwest cannot be considered independent from the larger global agricultural market – farmers in the U.S. make decisions about what to plant based on world-wide food prices and supplies. As with many sectors, additional research into the costs and benefits of adaptation and mitigation strategies will assist agricultural stakeholders in making decisions about short- and long-term management pathways.

Urban and Rural Community Planning

Communities both large and small make planning and infrastructure decisions that have far-reaching impacts on their ability to grow and adapt to changes in climate and society – and on their contributions to climate change. In many cases, cities are dealing with aging infrastructure that must be replaced or retrofitted in the near future; they do not have the luxury of long study periods or of waiting for "greener" technologies to become available. Because much of the built environment is owned by individuals and the private sector, communities must balance revision of building codes and zoning laws (e.g., to promote energy efficiency and "smart growth") with demands that new regulations not impose too great a cost burden for the owners of new or retrofitted buildings. In suburban and rural areas, migration outward from cities has led to changing demographics, new requirements for infrastructure (e.g., emergency services, roads, sewer systems, transportation lines back into cities), and changes in land use (from forests, fields, and farms to subdivisions, stores, and schools). Some of these areas do not have a centralized governance structure that manages zoning, transportation, and other aspects of planning on a local to regional basis.

Transportation

In urban areas, the population continues to both grow and spread outward. However, there are few transportation options for those living in the outer suburbs and in the fastest-growing counties vehicle miles traveled are still increasing. In Chicago, overall vehicle miles traveled per new acre of development is decreasing; the city and surrounding counties are looking into programs to reduce transportation emissions, which currently account for about 21% of total emissions. Approaches to emissions reduction already being pursued or in the planning stages include car-share programs, encouraging use of mass transit where available and expansion of mass transit to new areas, and developing new technologies. In addition to mitigation programs that focus on expanding low carbon transportation options, adaptation to the current and future impacts of climate change must also be considered. For example, more winter days in which temperature is near freezing (increasing the number of freeze-thaw cycles) may result in a greater number of potholes developing. In the summer, increased temperatures can lead to greater heat stress on roads and airport tarmac and to higher rates of fuel evaporation from airplanes. Because of the reliance on the Great Lakes and Mississippi River for long-haul transportation of raw materials and goods, the effects of climate change on water levels, storms, and ice cover are closely linked with transportation. Likewise, choices about transportation can also have important impacts and benefits for sectors such as ecosystems and public health.

Hydrology and Water Management

The lakes, rivers, and aquifers that supply the Midwest with water are governed by a patchwork of public and private entities that often consider water quantity and water quality separately from each other. A further complication arises from the international nature of the Great Lakes; a number of interstate and international treaties and compacts govern use aspects of water quantity and quality in the basin. As precipitation patterns change, demands placed on various water sources will also change; such changes may lead to disruptions in the local and regional water cycle (e.g., pumping groundwater into surface water systems, which in turn could lead to problems with water quality and land subsidence). Changing precipitation patterns will also put additional stress on water treatment facilities, sewers, and storm drains, which may not be sufficient to handle moving and treating runoff from stronger storm events. For example, the dramatic increase in rainfall intensity since 1995 (an increase of 300%), may lead to multiple flooding issues, particularly raw sewage entering the storm sewers during storm events and other storm water management issues. A critical problem is that engineering practices are generally based on past "design events" rather than taking current and projected trends into account. There are also significant emerging problems with transportation and flood control infrastructure, agricultural practices, soil and bank erosion, changing river morphology, and oxygen depletion due to warmer streams with impacts on stream and lake food webs.

NOAA/MODIS satellite image of the Great Lakes. Courtesy of Jeff Schmaltz.

Health

While it is apparent that there are links between climate change and health (*e.g.*, increased temperatures lead to higher incidences of heat-related illness and death, the effects of air pollution are compounded by higher temperatures and humidity, incidences of vector and water-borne diseases are likely to increase), more information about and research into the full range of impacts of climate change on human health are needed. Current climate change scenarios predict significant increases in the number of 90°F+ and 100°F+ days for cities like Chicago and Minneapolis; increased heavy storm events are likely to put even greater strain on water treatment plants, increasing the likelihood of untreated sewage releases into waterways. Health is closely linked to sectors such as transportation and recreation: levels of physical activity may increase or decrease (thus changing people's risks for diseases linked to sedentary lifestyles); people may spend more or less time outdoors (thus changing their exposure to disease vectors and air pollution). Also important to consider are the indirect effects of climate change on health. For example, as temperatures rise, calls to police about interpersonal violence also increase. Finally, the ability of local health departments to respond to the challenges of climate change may be constrained by insufficient budgets.

Indigenous Peoples

The Midwest has a number of indigenous communities and large areas of tribal lands. There are many potential mitigation opportunities that these communities may be considering, but in many cases these opportunities may also create negative environmental impacts – thus more information about trade-offs between options is needed. Representatives from various communities have been active in calling for reductions in greenhouse gases and

equitable methods for addressing climate change. It will be important for the Assessment process to consider both written and oral traditions and ways of knowing as a part of the engagement strategy.

Carbon Management and Mitigation

The Assessment efforts provide an opportunity for a broader framework for examining land and forest management and its impacts on the carbon cycle. For example, how does forest harvest rotation relate to other objectives, such as carbon sequestration? There are multiple linkages between water, agriculture, and energy that need to be explored in the context of carbon management and mitigation. Linkages and implications for other resources such as fisheries, food security, greenhouse gas emissions and biodiversity also need more attention.

Energy

Most midwestern states have identified renewable energy zones for further development of alternative energy capacity. For these zones in particular, and for the region in general, it will be important to identify what information can be provided to guide development (*e.g.*, wind potential). There are important legal questions tied to renewable energy development that may play a role in the ultimate selection of pathways, these include dealing with energy returned to the grid by customers' solar panels or windmills and legal challenges related to the placement of offshore wind facilities in the Great Lakes. Energy providers in the Midwest must be aware of energy production and demands across the United States and Canada, as the energy grid is linked across regions. The energy sector will also need information about how energy demands may change with changing climate, for example, if and how the demands for winter heating and summer cooling may stress the grid. Mitigation decisions in the energy sector, such as the use of carbon sequestration technologies, may have implications for other sectors; for example, choosing to retrofit a coal-fired power plant with carbon sequestration technology may prevent the realization of co-benefits such as reduced air pollution that may arise from replacing the plant with alternative energy sources.

Economics, Business, and Industry

Previous assessments have not included economic assessments; without economics as a part of the Assessment, it will be difficult to engage business and industry in any discussions – and it will be important to engage representatives from these sectors in assessment-related discussions. For example, cost is often cited as a major barrier to adaptation; doing a better job of assessing the costs and benefits of adaptation and mitigation would be helpful. Thus we must consider how economics and climate might be brought together in decision processes in order to avoid unpleasant surprises and maladaptive responses. In the Midwest, the manufacturing industry (*e.g.*, automobiles, farm equipment) plays a major role in the regional economy; the products of this industry are major contributors to emissions and a target for new technology development. In land use decision making, while land use is an important integrator of adaptation and mitigation themes, it is ultimately economic considerations that drive most land use decisions. Thus altering financial incentives may be the most important adaptation option available.

Recreation and Tourism

Many communities and individuals in the Midwest rely on income from recreation and tourism; it is important to understand how climate change will impact seasonal activities such as ice fishing, skiing, and beach-recreation. In addition to the ecosystems impacts cited above (*e.g.*, changes in fish distribution, invasive species, algae growth, and coastal erosion), climate change is likely to affect both the built infrastructure (*e.g.*, docks and water levels at marinas) and natural systems (*e.g.*, water quality at beaches, prevalence of pests such as mosquitoes) important to recreation and tourism. Furthermore, it may make tourism to specific destinations more or less attractive as temperatures and precipitation patterns change. For states that rely on hunting / fishing permits and park entrance fees to support much of their resource management budgets, significant changes in the distribution and abundance of target species or impacts on park amenities may severely challenge their ability to maintain and manage cultural and natural resources.

Cross-cutting Themes
Climate Change as a Priority

The recession and high unemployment have changed people's willingness to see climate change as a priority issue, but it is still possible to incorporate climate change considerations into virtually all of the region's planning and implementation activities and to use a climate change lens on decision making more generally. For example, the Chicago Climate Task Force is looking at foreclosure issues and the connection between climate, utility bills, and inability to pay the mortgage. A key lesson from this is that even if the motivation to resolve an issue came originally from a desire to limit vulnerability to climate change, there are ways to integrate these considerations into everyday decisions in ways that promote other co-benefits.

Cross-Sectoral Interactions and Life-Cycle Analysis

There is a need to look at intersections between the sectors in order to better assess how decisions in one sector can exacerbate or ameliorate the impacts of climate change and our ability to respond in related sectors. Taking such a systems-based approach is important for public policy. For example, there have been well-publicized concerns that certain biofuels programs (*e.g.*, corn ethanol) may actually increase emissions because of associated changes in water use, transportation, nitrogen- based fertilizer use, *etc*. For example, a recent life-cycle analysis of biofuels has shown an increase in particulate air pollution created by burning corn-based ethanol vs. gasoline.

Documenting Change

Some climate-related information needs include updated quantitative assessments of components of the hydrologic cycle (*e.g.*, evapotranspiration rates, soil moisture), documentation of current and anticipated land use changes, and ability to predict changes that link across sectors and time scales. Seasonal changes, such as changes in length of growing season, need to be evaluated in the context of increasing intensity of rainfall and the impacts of crossing thresholds at small spatial scales (*e.g.*, localized extinctions) need to be aggregated (*e.g.*, soil, water and biochemical systems changes and changes in the function of wetlands).

Sources of Vulnerability

Some of the sources of vulnerability that participants identified are similar to lists that would be generated in other places across the country, for example: lack of information about system dynamics, difficulties in decision making in the context of a non-stationary climate system, environmental justice concerns, aging infrastructure, and inadequate institutions and policies to address resource issues. Participants also expressed concern about threshold and feedback issues; a need to appreciate the interdependence of all sectors (*e.g.*, extreme heat and implications for electricity use and cooling load); and a lack of resources available to invest in mitigation and adaptation, especially in current economic conditions.

There are many external pressures and economic vulnerabilities that cross regional boundaries, including national security considerations and geopo-

Box 2: Suggestions for Specific Adaptation and Mitigation Options†

The following is a partial list of ideas from workshop participants:
- Drainage tiles to decrease the impact of runoff and flooding
- Tree planting
- Green roof technology, cool-roof rebates
- Carbon sequestration
- Permeable pavements
- Valve closing units on the storm sewers on the bay side of the systems
- Regional climate registries and carbon exchange programs—no-till farming, payments for sequestration (essentially, renting a space to store carbon – if the carbon is released after a period of time, then the person who bought the credits needs to buy more elsewhere)
- Channeling of resources to neighborhoods that are most vulnerable (*e.g.*, Green Impact Zone in Kansas City)

litical/trade implications of resource distribution and use. In the Midwest, the automobile and agricultural industries are particularly affected by these factors.

Identifying and Implementing Adaptation and Mitigation Options

Adaptation and mitigation decisions are already being implemented by early adopters and new options are under consideration. Participants highlighted a number of sectors in which adaptation and mitigation are already taking place, sectors ripe for development of new adaptation and mitigation options, and the challenges and barriers to developing and implementing adaptation and mitigation approaches. They also highlighted specific options that should be pursued in the Midwest (Box 2).

What is Already Being Done?

Climate change adaptation and mitigation are occurring in the Midwest and around the nation. In some cases, these activities are deliberately linked to climate change, while in others their benefits for adaptation or mitigation are secondary to another goal. Participants highlighted activities in a number of sectors, as described below.

Agriculture

Farmers have been adapting to climate change in many ways. Some practices, including soil conservation and conservation tillage measures, are long-established. Because many farmers can adjust operations on a seasonal to annual basis, they are planting crops earlier in response to earlier springs and use longer-season, drought-tolerant hybrids. In some cases, it is possible to cultivate more plants per acre. Farmers have increased use of fungicides and other chemicals in response to increased disease and pests. Additional drainage tiles are being installed to accommodate an increase in intensity of rain and runoff.

Water Management and Riparian Ecosystems

New storm water and flood management systems can be built with an eye toward increased variability in precipitation events and existing systems can be retrofitted or renovated to better handle extreme events; both Chicago and Milwaukee are pursuing such measures. Riparian restoration projects add capacity for storm water management and trap sediment before it can wash downstream. In St. Louis, storm water rain gardens are growing in popularity; such gardens attenuate runoff to city sewer systems and provide aesthetic or other benefits. In the west, San Francisco has implemented storm water and sea level rise planning. On a national scale, recent and planned revisions of the FEMA flood maps to better account for changing floodplain boundaries and flood regimes is one way that adaptation has already begun.

Built Environment
The city of Chicago has taken an active role in implementing adaptation and mitigation practices as a part of the Chicago Climate Action Plan process, including installing a green roof project on City Hall and making use of permeable pavement. During summer heat waves, many cities operate "cooling centers" to accommodate people who do not have air conditioning or whose electricity has failed. Other cities, including New York and Seattle, are also leading the way in developing climate change and sustainability plans. Other adaptation and mitigation measures are being implemented as a part of green building projects – both in new buildings and retrofitting existing buildings.

Energy
In the Midwest, CO_2 emissions come primarily from the energy (40%) and transportation (30%) sectors. According to the Energy Information Administration (EIA), demand for energy/electricity in the Midwest will remain flat through 2030 as a result of changes in manufacturing, increasing efficiency in buildings, and implementation of existing or new state energy efficiency guidelines and renewable energy standards (such standards exist in all midwestern states except Indiana). There may even be a 12 to 15 percent decrease in energy demand due to increasing efficiency. A number of governors in the Midwest signed onto the Midwestern Greenhouse Gas Reduction Accord; however, related activities of the Midwest Governors' Association (e.g., those associated with the Midwest Governors Energy Security and Climate Stewardship Platform) are now on hold as they wait to see what happens at the federal level. Implementing new energy efficiency standards should not be stopped by arguments of a poor economy – in many cases, these standards will save people money. These standards can also be used to change the mix of energy sources. For example, by holding nuclear energy and natural gas consumption about even, or perhaps slightly increasing the use of natural gas, and implementing state energy efficiency regulations, the market share of coal in the energy market could be decreased from approximately 70% in 2010 to approximately 40% in 2020.

The Union of Concerned Scientists has set forth a National Blueprint for a Clean Energy Economy which suggests that by using a combination of strategies (energy efficiency, alternative methods of transportation, etc.) the United States can reduce its global warming emissions to 26% below 2005 levels by 2020 and 56% below 2005 levels by 2030.

Transportation
Accessible and affordable public transportation is essential for meeting pollution / emissions reduction goals and provides a number of co-benefits tied to sectors such as public health. On the potential actions side, Chicago is a potential hub for a new national high-speed rail network. While increasing fuel usage standards and developing alternative vehicle fuels will play an important role in mitigation, participants cited the quick development and acceptance of corn ethanol (and later findings that it displaced food crops and could actually contribute to higher emissions overall) as a cautionary tale showing that more analysis of alternative fuels may be needed before they are brought to market.

What Research and Tools are Still Needed?
Participants identified a number of research areas and decision support tools that are needed to fill gaps in knowledge and increase capacity for decision making about adaptation and mitigation. One important distinction that must be made is between what users want *vs.* what the users actually need – many wants are either highly costly or currently not possible. When information producers and information users engage in a dialogue with each other, it is possible to help both sides better understand the capabilities and gaps – and to define needs in a productive way.

Synthesis, Reanalysis, and Interpretation of Existing Data Sets

Suggestions for use of current data included analysis of dynamics of temperature and precipitation patterns, as well as examination of thresholds, feedbacks, and nonlinearities.

"Right-Scaling" Climate Model Outputs

A main focus was on the need for "right-scaling" as opposed to downscaling of climate models. It is important to know with what scale the user of the data is concerned in order to deliver model outputs which meet their needs – and to suggest ways to get around current scale limitations. For example, the Water Utility Climate Alliance (WUCA) recommended specific improvements in climate models for the water utility sector as well as noting ways that managers are "working around" current limitations. Similar analyses could be done for other sectors and at other scales. Additionally, decision makers need multiple time scales of climate projections – one or a few years for operating decisions, decade scale to understand variability, and long-term (50 year scale) for infrastructure decisions.

Long-Term Monitoring

An investment in long-term monitoring of and research on impacts would be useful for evaluating the success of adaptation and mitigation activities and for adjusting these activities to maximize benefits. Information regarding consequences of land-use change would also be extremely useful to many users.

Vulnerabilities, Thesholds, and Feedbacks

There is a need for better assessment of vulnerability of critical systems. Physical and ecological studies can help us identify thresholds for change in ecosystems and characterize feedbacks between climate, ecosystems, and human systems. Social and economic information is necessary to identify vulnerable areas and target or prioritize resources. Evaluation studies and economic studies of specific climate change impacts and their societal values would also be useful for prioritizing adaptation and mitigation options. An example would be to quantify the co-benefits of health as a result of adaptation and mitigation activities.

Extreme Events

A recurring request was for a better understanding of extreme events and long-term drought. These are currently poorly parameterized and have greater impacts than the climatic means.

Economic and Social Information

In addition to needing information about vulnerability, as noted above, decision makers need information about long-term environmental and fiscal costs and benefits of adaptation and mitigation strategies. Information about how people value climate change and resources impacted by climate change will be useful in constructing better methods for economic analyses.

Communication Tools

We still are relatively unskilled in communicating about climate change to a broad public and additional research on and examples of successful communication strategies are needed. This includes learning about how to communicate with various user groups using language and analyses that are familiar to that community. We also have much work to do in building networks to deliver information to the translators and communicators most able to connect to various regions and sectors, including state climatologists, extension specialists, and other trusted sources and opinion leaders.

Decision Support, Discussion Support, and Co-Production of Knowledge

Technical training for decision makers would help them to make better use of the data available to them. However, because decision makers are used to incorporating new information in their own decision processes, we could be thinking more broadly about providing "discussion support" as well as decisions related to adaptation and mitigation. Understanding the sorts of decisions that people are making can lead to a better understanding of how to best provide science and information. It can also lead toward "co-production" of knowledge or "participatory research," in which the stakeholders help to formulate the problem statements and questions in such a way that the resulting research and information is better able to feed into decision making processes.

Performance Metrics

Once decisions are made, we need to be able to assess how well the selected adaptation and mitigation activities are performing. Thus we need to undertake research on performance metrics and evaluation methods, and on how to incorporate these metrics into adaptive management. In addition, other fields may have metrics that are useful; for example, real estate uses a "walk score" to rate the proximity of a house to nearby stores and transportation options and the accessibility of safe walking paths to these amenities. Such a score could be repurposed to rate the amount of emissions reduced or health benefits gained by regular use of these walking routes instead of car transportation.

Selecting and Implementing Options
Once decision makers have the information and tools they require to select and implement adaptation and mitigation measures, participants noted that there are several additional considerations that play a role in the decision making process.

"Rightscaling" Implementation
Adaptation and mitigation may be best accomplished at the municipal to watershed scale, and where climate change can be integrated into existing policies. Adaptation and mitigation must also be implemented at the appropriate time and fiscal scales, and thus more information about the short- and long-term benefits vs. near-term costs will be helpful for decision making. For the National Assessment, a lingering question is: how do we do an assessment at the local scale and ramp it up to regional and national levels?

"No Regrets" Measures
It would make sense to implement many of the adaptation and mitigation options discussed at the workshop because they are beneficial for the environment and the economy even without taking into account their ability to mitigate CO_2 emissions or adapt to climate change. Efforts to achieve energy efficiency serve multiple purposes and achieve multiple goals—saving money is good for the economy, reducing air pollution is good for public health, and retrofits for buildings create jobs and are smart for building management.

Promoting Adaptation and Mitigation
RISAs, Sea Grant, ICLEI, and the Great Lakes Initiative are actively promoting adaptation and mitigation activities. The USDA is hiring climate specialists (by state) as a reaction to the increasing number of climate-related questions they are getting. Lending institutions that advise farmers are encouraging them to get crop insurance in order to protect against lost revenue.

A Systems Approach
Using a systems approach – recognizing dynamic linkages across regions and sectors, and developing adaptation/mitigation approaches that facilitate integrative thinking and planning – can help reveal both unintended consequences and co-benefits of climate action. One planning method that uses this approach is integrated watershed management. Furthermore, we should go beyond describing impacts and evaluate future conditions under a range of scenarios and provide technical guidance on how to implement options.

Barriers and Challenges
There are many barriers to adaptation, including many institutional issues and cross-jurisdictional issues. While many of these barriers are not unique to the Midwest, the particular economic and social conditions of the Midwest combine with climate change to present unique challenges. Although there are some successful examples of how to overcome such issues, such as a regional effort (Green Impact Zone) in Kansas City, Missouri, it is difficult to develop a long-standing, cooperative, regional approach. Cultural norms and lifestyle choices are a major barrier to progress, as is lack of flexibility in institutional infrastructure. Workshop participants identified critical information, social and behavioral, and institutional and fiscal barriers and challenges that must be dealt with as a part of assessing and addressing climate change, as well as a number of specific communication and framing issues (Box 3).

Information
In many places, "credible" information and information sources (including the people delivering the information) are still needed. Even when information comes from a trusted source, it may be too technical or in a format that is inaccessible to the users, and so additional translators and technical specialists will be needed to assist in making information available.

Social and Behavioral
Climate change is still a contentious issue in the public sphere, thus linking action on other issues to action on climate change may ultimately invite criticism and reduce support for action. For example, linking the Great Lakes Restoration Initiative to climate change (e.g., restoration can also contribute to resiliency to climate change impacts) brought about opposition in the media. Many potential adaptation and mitigation options will require people to alter their behaviors; more research on how people react to change and ways to encourage behavior changes will be critical for successfully implementation of these options.

Another significant barrier is a lack of leadership, especially at the national level, and the lack of a vision for and integration of adaptation and mitigation at regional and smaller scales. In addition, there are different perceptions of the urgency of this issue due to differential impacts across regions; for example, some impacts (e.g., water stresses) are evident in the Southwest and less evident in the Midwest. This has complicated efforts to make a case for coordinated national climate policy action. There are also incongruities of scale (temporal and spatial) between, for example, climate change, electoral politics, jurisdictional boundaries, and budget cycles.

Institutional and Fiscal

Many of the constraints that participants note are fiscal and institutional. Implementing adaptation and

Box 3. Communication and Framing Issues

Participants identified a number of communications and framing issues that are barriers to action on climate change, including:

- Communicating impacts and climate change information in the context of a polarized public with set perspectives.
- Communicating uncertainty, including the sources of uncertainty and guidance on how to deal with this uncertainty.
- Need for probabilistic climate information (recognizing uncertainty but providing some confidence with regard to certain data points).
- Avoiding overpromising what we can provide.
- Guidance on how to describe and respond to extreme events (high impact, low probability events) and how to talk about them.
- Market research on what information specific stakeholders require so that workshops and materials can be crafted to fit their needs.
- Assessing how to get information to specific groups of people who can further distribute it (e.g., connecting to networks such as state climatologists).
- Guidance on how to help people understand the links between past and current climate trends (e.g., current climate is great for agriculture in Iowa but that does not mean that climate change will always bring better agricultural conditions).
- Considering needs vs. wants - many stakeholders want information that is not available or ask questions that are not answerable.
- Need for much more research on human behavior, particularly risk perception and how that motivates decisions to act or not act.
- Need to reframe responses to climate change in a more positive light, capitalizing on an opportunity to take a more holistic, integrated and sustainable view of planning problems.

mitigation measures will require cross-jurisdictional cooperation. In some places, the federal, state, and local policies may be inconsistent with each other, making it difficult to select a set of guiding criteria; there may also be legal barriers limiting the ways in which certain options are implemented. Securing the fiscal and administrative resources (including staff) to research, implement, and continue to evaluate and adjust adaptation and mitigation measures is a high barrier for some jurisdictions to overcome, especially when other, more immediate needs loom large.

Building a Next-Generation Regional Engagement and Assessment Process

A decade removed from the conclusion of the first National Assessment, the number of people interested in participating in current and future rounds of assessment has grown by one to several orders of magnitude. These people – and the myriad entities they represent – are part of the constantly expanding stakeholder community that the National Assessment process must be sure to engage. Workshop participants identified a number of issues related to building an assessment process that is responsive to the questions and needs of individual stakeholders, to the various regions and sectors on which are the focus of assessment, and to the spirit of the legislative mandate which established the National Assessment.

What are the Key Considerations for Stakeholder Engagement?

As people and organizations look for more and better information about climate change and its impacts, the National Assessment must be structured in such a way that a wide variety of stakeholders can access its process, information, and products. The following themes emerged from discussions about how to ensure that the Assessment moves toward meaningful stakeholder engagement.

Engagement is Essential

The First National Assessment involved an unprecedented level of effort – and yet ultimately fell short of its goal for rich regional engagement. Because responsibility for regional and sectoral chapters was distributed among agencies with minimal central guidance, the approach to engagement and assessment was uneven and in many cases too hands-off. In addition, the report was released at a difficult time politically, near the transition between administrations. Because there had not been strong engagement, stakeholders did not feel ownership of the Assessment and ultimately it fell victim to partisan attack. Building a stronger engagement process, including encouraging more conversations between stakeholders, scientists, and assessment practitioners (hopefully leading to self-sustaining dialogues), will likely result in greater stakeholder buy-in to the validity and importance of the Assessment process and its products and thus to a stronger support base for the Assessment.

Begin with the End in Mind
In order to effectively engage stakeholders, the process must begin with clearly-defined goals for both the engagement process and for the National Assessment as a whole. Having a set of goals for the engagement itself will be necessary in order to explain to stakeholders why we are inviting them into the process and how they might benefit. This set of goals will also help the organizers avoid being caught up in engagement for the sake of engagement and exhausting themselves and the stakeholders; put differently, the "why" of the Assessment will help organizers to better define "who" should be involved and "how" to involve them.

Build on Existing Networks
Federal extension networks, such as Sea Grant Extension and Agricultural Extension, have a long history of engaging stakeholders in the places where they live and work and have much expertise to offer in designing and carrying out an engagement process for the National Assessment. Federal programs (e.g., NOAA's Regional Integrated Sciences and Assessments (RISA) program and Sectoral Application Research Program (SARP), NSF's Decision Making Under Uncertainty (DMUU) program, NASA's Applied Sciences Program) have supported many research and pilot projects that provide important lessons about engagement and have established networks of researchers and managers. The Assessment process should also work with existing networks of NGOs (e.g., Healing Our Waters, ICLEI, ASHRE), professional organizations (e.g., state climatologists, broadcast meteorologists, farmers unions), and groups such as local Chambers of Commerce and Rotary Clubs to identify "opinion leaders" and other key stakeholders to target for engagement. These networks will also serve as pathways to disseminate information coming out of the Assessment process to a larger community.

Engagement Must be Broad
The number of people interested in, or potentially interested in, climate change continues to grow as the outputs from natural and social science studies of climate change impacts help us to understand the breadth of the problem. At the same time, climate change may continue to stay toward the bottom of stakeholders' agendas as more immediate issues are dealt with first. Despite this, the demand for climate information is there, and the engagement process will help sustain a dialogue aimed at defining the questions and needs of stakeholders that the Assessment should address. In addition, stakeholders desire a range of engagement methods, ranging from in-person discussions to web-based content and social media; the engagement process will need to consider how it will draw on the various tools and methods that are available in order to maximize meaningful participation. The idea that engagement must be broad also applies to the Federal agencies involved in leading the engagement and Assessment processes: while science agencies have taken the lead in past processes, it is necessary and appropriate for a wider range of agencies to participate as both conveners and stakeholders, as they may have specific skills and expertise important to the process.

Engagement Must be Sustained
The purpose of stakeholder engagement cannot be only to frame an assessment and then, one or several years later, deliver a product. Stakeholders must be regarded as experts with something of value to contribute throughout the process so that they are able to shape the process and its products. Engagement is time-consuming, as it requires multiple interactions and each interaction requires adequate time for a dialogue; planning for sufficient time and resources to support an extended engagement process is critical.

Engagement is Non-Linear
There are already many successful efforts underway at local, regional, and national levels. The challenge will be to link these efforts and build on them. As the process engages new stakeholders, new questions are likely to arise and new methods of engagement and analysis may be required. Both information users and information producers will need to learn new vocabulary and work together to define a common language.

Focus on "Discussion Support"

We cannot know all of the decisions that an assessment might be asked to inform at the start of the process. Instead, the engagement process will provide the space for discussion support, bringing together stakeholders with scientists and other subject matter experts in a way that will help frame the goals and questions of the Assessment and will build capacity for better understanding and using the information and products coming out of the Assessment process. Creating this discussion space will also require entraining "translators" who can help ensure that conversations at the science-decision making interface are taking place in a way that all parties are able to understand. In some cases, these translators may come from fields that have not been deeply engaged in the past (e.g., the cultural sector, such as musicians and authors).

What are the Key Considerations for Regional and Sectoral Assessment Processes?

Many participants acknowledged that a top-down, highly-centralized Assessment process is not likely to be viewed as legitimate. Instead, the National Assessment must be carried out via a "distributed assessment system"[2] which empowers regions and sectors to define a process that fits their needs and strengths while providing a coordinated and collaborative framework that integrates findings and provides support at a national level. The following themes emerged from discussions about how to create such a distributed assessment process for the National Assessment.

Assessment is Both a Process and a Set of Products

Participants noted that the National Assessment is more than a single report that comes out every four years. Instead, the Assessment should be viewed as a process for bringing together scientists, decision makers, and stakeholders to frame the risks and vulnerabilities related to climate change in such a way that stakeholders are able to use this information to support their decisions about preparing for and responding to climate change through adaptation and mitigation. The Assessment must be designed in such a way that it provides regular reports and products which incorporate the best science available while also being nimble enough the address emerging issues and providing people with a mechanism for recognizing their climate-related needs, finding the information they need (and more importantly, the people who have the capabilities to help them use this information), and making their voices heard in the ongoing process.

Shared National and Regional Leadership

The parties responsible for planning and carrying out the regional and sectoral process need to be of that region or sector (regional granularity), but they must be connected to and receive support from a nationally-coordinated Assessment office (national coherence). Locating major leadership responsibilities at the regional and sectoral levels will help ensure that the Assessment addresses questions and issues relevant to that particular community and increases the likelihood that stakeholders will engage because the conveners of the process will be known to them. Leaders at the regional and sectoral level can also help to identify and entrain existing initiatives that may serve as examples or early "test beds" for Assessment-related activities. Support from the national level can help facilitate better stakeholder engagement processes by demonstrating a commitment to engagement, link assessment practitioners across regions and provide channels for communication, provide oversight to ensure that the Assessment is proceeding in all areas, and identify common needs and resources.

Boundaries are Mutable

While a regional and sectoral approach to assessing the impacts of climate change is generally appropriate, the process must allow for boundaries to be fuzzy and for links across the various pieces of the Assessment. For some issues, it may be desirable to convene a "transboundary" team that brings together specific expertise and perspectives. In other cases, the areas of focus must be allowed to arise from conversations among participants. For example, the UK Climate Impacts Program organized around sectors and vocabulary that were most relevant to people's everyday lives, eventually producing reports on topics such as gardening, housing, food, floods, transportation in a particular city, and retail.

Co-Production of Knowledge

Engaging a wide variety of stakeholders in co-producing the knowledge generated within the Assessment (i.e., as part of the author teams) will likely lead to increased acceptance of the Assessment, as those involved will take ownership of the pieces in

---

[2] Cash, D.W. 2000. Distributed assessment systems: an emerging paradigm of research, assessment and decision-making for environmental change. Global Environmental Change 10: 241-244.

which they participated. Moving beyond the realm of government and academia – involving private citizens, NGOs, and industry – can be challenging, especially as it will require much effort to identify the "right" level of people to engage. However, engaging in this co-production of knowledge will ultimately build capacity within the community and will result in a stronger process overall.

Diversity AND Consistency

In the First National Assessment, individual agencies took on responsibility for various pieces of the Assessment. This arrangement led to inconsistencies in the level of effort and the process used in different regions and sectors. While each region and sector will have its own unique set of climate change-related impacts and vulnerabilities, there are many common problems. Therefore, the Assessment process must provide a mechanism for sharing knowledge and approaches across regions and sectors in such a way that practitioners can learn from each other and produce regionally-relevant results while also allowing for findings to be aggregated at a national level.

Adequate Resources

In order for the Assessment process to be successful, it must be adequately resourced. This means providing for a functional and sufficiently-staffed office at the national level that can coordinate across regions and sectors, regional offices that are empowered to convene stakeholder engagement programs and to do the bulk of the work of the Assessment (including supporting sector-based efforts), and additional resources for technical support and new areas of interest.

Next Steps

The organizers of the Assessment process have much to consider as they work toward a framework for implementing the next National Assessment and an ongoing, sustainable process that will continue into the future. Here we present a summary of concerns raised at the construction of the workshop itself and key questions that must be answered as a part of the continuing development of the National Climate Assessment process.

Summary of concerns regarding the Midwest workshop format and effectiveness:

Perhaps the strongest message articulated about this workshop by the participants relates to engagement of private sector stakeholders. Although there were

many regional representatives from universities and federal agencies, the number of private citizens, businesses, and local governments represented was relatively small. This was attributed to several problems in designing the workshop itself:

- We were not clear enough about what we were trying to accomplish in this workshop, *e.g.* was this the major opportunity for stakeholders to define their information needs, or was it really just an opportunity to provide regional input into the design of the National Adaptation Strategy?

- The private and public sector local stakeholders who were invited in many cases chose not to attend. This may be in part due to a lack of clarity of "what was in it for them." In some cases, they also did not have enough lead time to get the meeting on their calendars.

- USGCRP recently completed a series of 22 "listening sessions" around the country in which stakeholders were asked similar questions.

- There is now a very large community of people engaged in the "provider" side of the climate information equation who are extremely anxious to be engaged in the development of the next National Assessment. This resulted in a large number of government and academic climate community people wanting very much to be at this "kickoff" workshop.

- Local opinion leaders were not involved in the development of the workshop concept from the beginning.

- The definition of "stakeholder" necessarily includes everyone who has a "stake," and that group of people is now several orders of magnitude bigger than it has been in the past.

Key questions raised by the participants that must be answered as a part of the continuing development of the National Climate Assessment process:

- *Who should lead the process? Who should be included in developing a plan?* The leaders in the Assessment must be thought leaders who are trusted locally and regionally.

- *How can we manage and coordinate across regions?* There must be a plan for coordination and engagement fairly early in the process which lays out issues and concerns for each region. This should be a living document that can be adapted as conversations evolve, but it is essential to have this as a guide.

- *What methods should be used for engagement?* We must use methods that will hold the attention of stakeholders, be respectful of the time that participants have given to the process, and sustain dialogues once they have begun. The Assessment process should leverage existing meetings and forums, capitalize on opportunities for engagement provided by major climate events and key policy initiatives, and learn from past efforts to improve future success.

- *How can we keep stakeholders engaged?* Stakeholders often require an incentive to stay engaged – they must see the value of their involvement in the process. The process must include regular evaluation to ensure that engagement is working and should adapt when necessary.

- *Who are the decision makers?* Many types of stakeholders need climate change information for their decision processes. Some of these people have participated in past assessments and others have not.

- *What decisions are we trying to inform or support?* We do not yet have a full picture of the types of decisions that stakeholders expect the Assessment to support. Generally, the process must provide products and services that are applicable to a wide array of decisions, but more engagement is needed to better frame the issues and decisions of interest.

- *What types of tools and support services need to be developed?* Many tools already exist, but we do not have a good idea of the existing tools and services – thus we must do some sort of capability mapping to better understand what is available. In the end, most stakeholders still want connections to real people on the other end of the tool, so user support is essential.

- *What platforms should be supported?* There is no one "winning" platform or access point for climate science and information – stakeholders discover and use parts of the Assessment in different ways, including through web-based tools, social networks, and in-person discussions.

- *How can we ensure that the regional process is robust and ongoing?* How do we encourage long-term ownership of the process and its products? Climate must be embedded into all decision making cultures, which will require engaging with myriad stakeholders and helping them to understand the value of the Assessment process. In the Great Lakes area, it will be especially important to ensure that the process includes links to international activities.

- *Where will the resources for the Assessment come from?* A dedicated resource stream is essential to the success of the Assessment process.

Lake Superior, ©iStockphotos.com/EricFoltz

Appendix A: Agenda

Monday, February 22nd

4:00 – 4:30 pm	**Welcome from Chicago Hosts** (D. Wuebbles – University of Illinois, Joyce Coffee – City of Chicago)
4:30 – 4:45	**Overview and Charge for this meeting** (J. Melillo)
4:45 – 5:30	**The Impacts of Climate Change on the U.S.**
	• An overview of the changing climate (T. Karl)
	• Our climate future (in the Midwest) (D. Wuebbles)
	• Impacts (T. Janetos)
5:30 – 6:00	**Questions**
6:00 – 6:30	**Reception**
6:30	**Dinner: welcome by T. Karl and keynote**
	Shere Abbott, Associate Director for Energy and Environment, Office of Science and Technology Policy (invited)

Tuesday, February 23rd

7:30 – 8:30 am	**Continental Breakfast**
8:30 – 9:00 for	**Overview of the charge, agenda and intent of first day** (K. Jacobs - opportunity feedback and questions on agenda and plan for the workshop)
9:00 – 10:15	**Panel Discussion: Critical sectoral and regional issues** (D. Wuebbles, introductions and chair)
	• Scott Bernstein, Center for Neighborhood Technology, Chicago: Land use and transportation
	• Jonathan Patz, University of Wisconsin: Health
	• Knute Nadelhoffer, University of Michigan: Forest ecology
	• Efi Foufoula-Georgiou, University of Minnesota: Water
	• Charlie Walthall, USDA Agricultural Research Service: Agriculture
10:15 – 10:30	**Expectations/instructions for breakout groups** – K. Jacobs
10:30 – 11:00	**Break**
11:00 – 12:30	**Breakout sessions on identifying sectoral, regional and cross-sectoral issues and questions in the Midwest**
	Facilitators: K. Jacobs, J. Buizer, L. Carter, E. Shea
	Rapporteurs: F. Niepold, P. Runci, A. Waple, E. Cloyd
12:30 – 1:45	**Lunch** – Mark Howden, CSIRO Adaptation Flagship, Australia
1:45 – 2:45	**Panel discussion on identifying adaptation and mitigation options to address key issues that have been identified** (M. Gade, Chair)
	• Gene Takle, University of Iowa, Agriculture
	• Howard Learner, Environmental Law and Policy Center, Public Policy
	• Ron Burke, Union of Concerned Scientists
	• Don Scavia, University of Michigan, Great Lakes Issues
2:45 – 4:00 key	**Breakout sessions on identifying adaptation and mitigation options to address issues**
	Facilitators: P. Runci, D. Brown, T. Janetos, J. Melillo
	Rapporteurs: N. Engle, F. Laurier, J. Austin, N. Gardiner
4:00 – 4:30	**Break**

4:30 – 5:45	**B**reakout sessions on identifying <u>science and information needs for adaptation/ mitigation decisions</u> for key sectors and interest groups
	Facilitators: K. Jacobs, J. Buizer, L. Carter, E. Shea
	Rapporteurs: N. Gardiner, D. Ferguson, A. Waple, E. Cloyd
5:45	**Dinner on your own**

Wednesday, February 24th

7:30 – 8:30	**Continental Breakfast**
8:30 – 9:10	**Summary of previous day's findings from facilitators** (T. Janetos)
9:10 – 10:30	**Panel Discussion: How to build a next-generation regional engagement and deci sion support process** (Kathy Jacobs, Chair)
	• Richard Moss, University of Maryland
	• Tom Wilbanks, Oak Ridge National Laboratory
	• Diana Liverman, University of Arizona
	• Ted Parson, University of Michigan
10:30 – 10:50	**Break**
10:50 – 12:00	**Breakout sessions on next-generation regional engagement and decision support ideas**
	Facilitators: P. Runci, N. Gardiner, T. Janetos, J. Melillo
	Rapporteurs: N. Engle, F. Laurier, F. Niepold, J. Austin
12:00 – 12:30	Report to Plenary from breakouts and discussion
12:30 – 1:30	Lunch and Next Steps (J. Melillo and Kathy Jacobs)

Appendix B: Participant List

Jimmy Adegoke
Professor
Environmental Studies Program
University of Missouri-Kansas City
420 R.H. Flarsheim Hall, 5110 Rockhill Rd.
Kansas City, MO 64110
Tel: 816-235-1334
Fax:
Email: adegokej@umkc.edu

Jeffrey A. Andresen
Associate Professor
and State Climatologist for Michigan
Department of Geography
Michigan State Climatologist Office
116 Geography Building
E. Lansing, MI 48824
Tel: 517-4324756
Fax: 517-432-1076
Email: andresen@msu.edu

Jim Angel
Director
Illinois State Climatologist Office
Illinois Water Survey, 2204 Griffith Drive
Champaign , IL 61820-7495
Tel: 217-333-0729
Fax:
Email: jimangel@illinois.edu

Thomas R. Armstrong
Senior Advisor for Climate Change
Office of the Deputy Secretary
U.S. Department of the Interior
1849 C St., NW
Washington, DC 20240
Tel: 202-208-6713; Cell - 703-304-0229
Fax: 202-208-1873
Email: tarmstrong@usgs.gov

Jennifer Austin
NOAA Communications & External Affairs
Washington, DC
Tel: 202-482-5757; Cell: 202-302-9047
Fax:
Email: jennifer.austin@noaa.gov

Kristen B. Averyt
Deputy Director
NOAA-CIRES, Western Water Assessment
University of Colorado at Boulder
NOAA Earth System Research Laboratory
325 Broadway R/PSD
Boulder, CO 80305-3328
Tel: 303-497-4344; Cell: 303-827-1059
Fax:
Email: kristen.averyt@noaa.gov

David Behar
Climate Program Director, San Francisco PUC
Staff Chair, Water Utility Climate Alliance
1145 Market St., 4th Fl.
San Francisco, CA 94103
Tel: 415-554-3221
Fax:
Email: dbehar@sfwater.org

Scott Bernstein
President
Center for Neighborhood Technology
2125 W. North Ave.
Chicago, IL 60647
Tel: 773-269-4035
Fax: 773-617-9503
Email: scott@cnt.org

Maria Blair
Deputy Associate Director
for Climate Change Adaptation
White House Council on Environmental Quality
(CEQ)
Executive Office of the President
722 Jackson Place
Washington, DC 20503
Tel: 202-456-1475
Fax: 202-456-6546
Email: mblair@ceq.eop.gov

Daniel Brown
Professor, School of Nat. Resources
and Environment
Director, Environmental Spatial Analysis Laboratory
University of Michigan
440 Church St., 3505 Dana Building
Ann Arbor, MI 48109-1041
Tel: 734-763-5803
Fax: 734-936-2195
Email: danbrown@umich.edu

Otis Brown
Rosenstiel School of Marine and
Atmospheric Sciences
University of Miami
4600 Rickenbacker Causeway
Miami, FL 33149-1031
Tel: 305-421-4000
Fax: 305-421-4711
Email: obrown@miami.edu

James L. Buizer
Executive Director
Strategic Institutional Advancement and Policy Advisor to the President
Arizona State University
P.O. Box 877705
Tempe, AZ 85287-7705
Tel: 480-965-6515
Fax: 480-965-0865
Email: buizer@asu.edu

Ron Burke
Midwest Office Director
Midwest Climate Campaign Director
Union of Concerned Scientists
1 N. LaSalle St., Ste. 1904
Chicago, IL 60602
Tel: 312-578-1750 x13
Fax:
Email: rburke@ucsusa.org

Lynne M. Carter
Director, Adaptation Network
Assoc. Director, RISA, LSU
Assoc. Director, Sustainability Agenda
Louisiana State University
E-333 Howe-Russell
Baton Rouge, LA 70803
Tel: 401-527-6058; Cell: 401-527-6058
Fax:
Email: lynne@srcc.lsu.edu

Macol M. Stewart Cerda
Silmaril Consulting
1523 W. Jackson Blvd.
Chicago, IL 60607
Tel: 312-404-4487
Fax:
Email: macolcerda@mac.com

Emily T. Cloyd
Carbon & Ecosystem Support Program Specialist
U.S. Global Change Research Program
1717 Pennsylvania Ave., NW, Ste. 250
Washington, DC 20006
Tel: 202-419-3484; Cell: 202-286-9642
Fax: 202-223-3065
Email: ecloyd@usgcrp.gov

Joyce Coffee
City of Chicago Department of Environment
30 N. LaSalle Ste. 2
Chicago, IL 60602
Tel: 312-742-0151
Fax: 312-744-6451
Email: joyce.coffee@cityofchicago.org

Kim Curtis
Program Director
Resource Media
325 Pacific Ave., 3rd Floor
San Francisco, CA 94111
Tel: 415-397-5000 x305
Fax: 415-397-5020
Email: kim@resource-media.org

Josh Darr
Meteorologist
Chesapeake Energy Corporation
100 North Riverside Plaza, Suite 2350
Chicago, IL 60606
Tel: 312-756-1805
Fax: 312-756-1807
Email: josh.darr@chk.com

Jon Davis
Meteorologist
Chesapeake Energy Corporation
100 North Riverside Plaza, Suite 2350
Chicago, IL 60606
Tel: 312-756-1805; Cell: 312-617-3943
Fax: 312-756-1807
Email: jon.davis@chk.com

Otto Doering
Professor of Agricultural Economics
Purdue University
1145 Krannert Building
West Lafayette, IN 47907-1145
Tel: 765-494-4226
Fax: 765-496
Email: doering@agecon.purdue.edu

Brenda Ekwurzel
Climate Scientist
Union of Concerned Scientists
1825 K St NW, Suite 800
Washington, DC 20006-1232
Tel: 202-331-5443
Fax: 202-223-6162
Email: bekwurzel@ucsusa.org

Nathan Engle
School of Natural Resources and Environment
University of Michigan
2209 W. Byron, #3
Chicago, IL 60618
Tel: 484-695-6185
Fax:
Email: nengle@umich.edu

Jack D. Fellows
Vice President for Corporate Affairs
Director, UCAR Community Programs
University Corporation for Atmospheric Research
P.O. Box 3000-FL4
Boulder, CO 80307-3000
Tel: 303-497-8655
Fax: 303-497-8638
Email: jfellows@ucar.edu

Edward Fenelon
Meteorologist In Charge (MIC)
NOAA National Weather Service - Chicago
333 W. University Dr.
Romeoville, IL 60446
Tel: 815-834-0600 x642
Fax: 815-834-0645
Email: Edward.fenelon@noaa.gov

Daniel Ferguson
Program Manager
Climate Assessment for the Southwest (CLIMAS)
Institute for the Environment
University of Arizona
P.O. Box 210156
Tucson, AZ 85719
Tel: 520-622-8918
Fax: 520-792-8795
Email: dferg@email.arizona.edu

Gabriel M. Filippelli
Department of Geology
Indiana University
Purdue University, Indianapolis
420 University Blvd, SL124
Indianapolis, IN 46202
Tel: 274-7484
Fax:
Email: gfilippe@iupui.edu

Efi Foufoula-Georgiou
Professor
Dept. of Civil Engineering
University of Minnesota
500 Pillsbury Drive S.E.
Minneapolis, MN 55455
Tel: 612-626-0369
Fax:
Email: efi@umn.edu

Peter Frumhoff
National Headquarters
Union of Concerned Scientists
2 Brattle Square
Cambridge, MA 02238
Tel: 617-301-8027
Fax: 617-864-9405
Email: pfrumhoff@ucsusa.org

Mary A. Gade
Gade Environmental Group, LLC
444 N. Michigan Ave., Suite 3600
Chicago, IL 60611
Tel: 608-669-8040
Fax:
Email: mary.gade@yahoo.com

Ned Gardiner
Climate Visualization Project Manager
NOAA Climate Program Office
NOAA National Climatic Data Center
151 Patton Ave., Rm. 557C
Asheville, NC 28801-5006
Tel:
Fax:
Email: ned.gardiner@noaa.gov

Kevin R. Gurney
Dept. of Earth and Atmospheric Sciences and
Dept. of Agronomy
Purdue University
CIVL 2277, 550 Stadium Mall Drive
West Lafayette, IN 47907-2051
Tel: 765-494-5982
Fax:
Email: kgurney@purdue.edu

Bethany Hale
Central Region Team Coordinator
National Oceanic and Atmospheric Administration
7220 NW 101st Terr.
Kansas City, MO 64153
Tel: 816-268-3133
Fax: 816-891-8362
Email: Bethany.a.hale@noaa.gov

Chris Hamilton
State Wildlife Biologist
Natural Resources Conservation Service
U.S. Dept. of Agriculture
Parkade Center
601 Business Loop 70 West
Columbia, MO 65203
Tel: 573-876-9416
Fax: 573-876-0913
Email: chris.hamilton@noaa.gov

Rebecca Held
University of Michigan
NOAA Great Lakes Environmental Research Lab.
4840 S. South Rd.
Ann Arbor, MI 48108
Tel: 734-741-2394
Fax:
Email: rebheld@umich.edu

Harry Hillaker
State Climatologist
Iowa Dept. of Agriculture & Land Stewardship
Wallace State Office Building
Des Moines, IA 50319
Tel: 515-281-8981
Fax:
Email: harry.hillaker@iowaagriculture.gov

Steve Hipskind
Division Chief, Earth Science Division
NASA Ames Research Center
MS 245-4
Moffett Field, CA 94035-1000
Tel: 650-604-5076; Cell: 650-279-1570
Fax: 650-604-3625
Email: steve.hipskind@noaa.gov

Mark Howden
Sustainable Ecosystems
CSIRO
Gungahlin Homestead
Bellenden Street
Crace, ACT 2911
AUSTRALIA
Tel: (61) 2-6242-1679
Fax:
Email: mark.howden@csiro.au

James P. Hurley
Assistant Director for Research and Outreach
University of Wisconsin Sea Grant Institute/
University of Wisconsin Water Resources Institute
Goodnight Hall
1975 Willow Dr.
Madison, WI 53706
Tel: 605-262-0905
Fax: 608-262-0591
Email: hurley@aqua.wisc.edu

Tom Iseman
Program Director for Water Policy
Western Governors' Association
1600 Broadway, Ste. 1700
Denver, CO 80202
Tel: 303-623-9378
Fax: 303-534-7309
Email: tiseman@westgov.org

Brian Jackson
Project Coordinator
Joint Office for Science Support
University Corporation for Atmospheric Research
P.O. Box 3000 - FL4, Rm 2334
Boulder, CO 80307-3000
Tel: 303-497-8663
Fax: 303-497-8633
Email: bjackson@ucar.edu

Katharine L. Jacobs
Assistant Director for Assessments and Adaptation
White House Office of Science and Technology
Policy
U.S. Global Change Research Program Office
1717 Pennsylvania Ave, NW, Ste. 250
Washington, DC 20006
Tel:
Fax:
Email: jacobsk@email.arizona.edu

Anthony C. Janetos
Director
Joint Global Change Research Institute
Pacific Northwest National Laboratory/
University of Maryland
5825 University Research Ct., Ste. 3500
Baltimore, MD 20740
Tel: 301-314-7843
Fax: 301-314-6719
Email: Anthony.janetos@pnl.gov

Alexa K. Jay
Research Associate, Climate Science Watch
Government Accountability Project
1612 K Street, NW Suite 1100
Washington, DC 20006
Tel: 206-849-5060
Fax:
Email: alexakjay@gmail.com

Lucinda Johnson
Director, Research Associate
Natural Resources Research Institute
5013 Miller Trunk Highway
Duluth, MN 55811
Tel: 218-720-4251
Fax:
Email: ljohnson@d.umn.edu

Thomas R. Karl
Director, National Climatic Data Center
Lead, NOAA Climate Services
NOAA National Climatic Data Center
Veach-Baley Federal Building
151 Patton Ave., Rm. 557C
Asheville, NC 28801-5006
Tel: 828-271-4476
Fax: 828-271-4246
Email: thomas.r.karl@noaa.gov

Caitlyn Kennedy
Science Writer
NOAA Climate Program Office
1315 East-West Highway, SSMC-3
Silver Spring, MD 20910
Tel: 301-734-1219; Cell: 301-706-0285
Email: caitlyn.kennedy@noaa.gov

Doug Kluck
Climate Services Program Leader
NOAA National Weather Service
7220 NW 101st Terrace
Kansas City, MO 64153
Tel: 816-268-3144
Fax: 816-891-7810
Email: doug.kluck@noaa.gov

Chester J. Koblinsky
NOAA Climate Goal Lead
Director, Climate Program Office
NOAA Climate Program Office
1315 E. West Highway, SSMC3
Rm. 12837 North
Silver Spring, MD 20910
Tel: 301-734-1233
Fax: 301-713-0515
Email: chester.j.koblinsky@noaa.gov

Nancy Laney
Community Volunteer
7163 Washington Ave.
St. Louis, MO 63130
Tel: 314-882-3202
Fax:
Email: laneynancy@yahoo.com

Fabien J.G. Laurier
NSTC Subcommittee on Global Change Research
U.S. Global Change Research Program
1717 Pennsylvania Ave, NW, Ste. 250
Washington, DC 20006
Tel: 202-419-3481; Cell: 202-288-2879
Fax:
Email: flaurier@usgcrp.gov

Howard A. Learner
Executive Director
Environmental Law and Policy Center
35 East Wacker Dr., Ste. 1300
Chicago, IL 60601
Tel: 312-673-6500
Fax:
Email: hlearner@elpc.org

Charles Lin
Environment Canada
4905 Dufferin Street
Toronto, ON
CANADA
Tel:
Fax:
Email: Charles.lin@ec.gc.ca

Diana Liverman
Institute of the Environment
The University of Arizona
Tucson, AZ 85721
Tel: 520-388-0190
Fax:
Email: liverman@u.arizona.edu

Brent Lofgren
NOAA Great Lakes Environmental Research Lab
4840 South State Rd.
Ann Arbor, MI 48108
Tel: 734-741-2383
Fax: 734-741-2055
Email: brent.lofgren@noaa.gov

Kerry Lofton
Office of the Director
Illinois Department of Agriculture
Chicago, IL
Tel: 312-814-4866; Cell: 312-802-5274
Fax: 312-814-4862
Email: Kerry.lofton@illinois.gov

Amy Luers
Senior Environmental Science Manager
Google.org
900 Alta Ave.
Mountain View, CA 94043
Tel: 415-736-1013
Fax:
Email: amyluers@google.com

Jeffrey C. Luvall
Global Hydrology and Climate Center
NASA -NSSTC
320 Sparkman Drive
Huntsville, AL 35805
Tel: 256-961-7886
Fax: 256-961-7788
Email: jluvall@nasa.gov

Ray McCormick
Vice-President
Indiana Association of Soil and Water Conservation
Districts (IASWCD)
225 S. East St., Suite 740
Indianapolis, IN 46202
Tel: 317-692-7519
Fax: 317-423-0756
Email: jennifer-boyle@iaswcd.org

Sabrina McCormick
Fellow, American Association for the
Advancement of Science
U.S. Environmental Protection Agency
George Washington University
US EPA (8601-P)
1200 Pennsylvania Ave., NW
Washington, DC 20460
Tel: 215-898-5456
Fax:
Email: sabrina.mccormick@gmail.com

Michael MacCracken
Chief Scientist for Climate Change Programs
Climate Institute
900 17th St., Ste. 700 (with Heinz Center)
Washington, DC 20006
Tel: 202-552-4723
Fax: 202-737-6410
Email: mmaccrac@comcast.net

James R. Mahoney
Environmental Consultant
18482 Lanier Island Square
Leesburg, VA 20176
Tel: 703-777-6333
Fax:
Email: mahoneyenv@aol.com

Lynn P. Maximuk
Director, Central Region
NOAA, National Weather Service
7220 NW 101st Terr. (CASC)
Kansas City, KS 64153
Tel: 816-268-3130
Fax: 816-861-8362
Email: lynn.maximuk@noaa.gov

Jerry Melillo
Director
Marine Biological Laboratory
7 MBL St.
Woods Hole, MA 02543
Tel: 508-289-7494
Fax: 508-457-1548
Email: jmelillo@mbl.edu

Edward L. Miles
Co-Director, Center for Science in the
Earth System
Director, Climate Impacts Group
JISAO, Dept. of Atmospheric Science
University of Washington
Box 355672
Seattle, WA 98195-5672
Tel: 206-685-1837 or 206-616-5348
Fax: 206-543-1417 or 206-616-5775
Email: edmiles@u.washington.edu

Christopher Miller
Program Manager, Climate Change Data and Detection
NOAA Climate Program Office
1100 Wayne Avenue, 12th Floor
Silver Spring, MD 20910
Tel: 301-734-1241
Fax: 301-713-0517
Email: Christopher.d.miller@noaa.gov

Kathryn Moran
Office of Science and Technology Policy
Executive Office of the President
New Executive Office Building
725 17th St., NW, Ste. 7217 (5th floor)
Washington, DC 20038
Tel:
Fax:
Email: kathryn_moran@ostp.eop.gov

Richard H. Moss
Senior Staff Research Scientist
Joint Global Change Research Institute
University of Maryland
5825 University Research Ct., Ste. 3500
College Park, MD 20740
Tel: 301-314-6711; Cell: 202-468-5441
Fax:
Email: rhm@pnl.gov

Michael Murray
Staff Scientist
National Wildlife Federation
Great Lakes Regional Center
213 West Liberty St., Suite 200
Ann Arbor, MI 48104-1398
Tel: 734-887-7110
Fax: 734-887-7199
Email: murray@nwt.org

Knute Nadelhoffer
Director, University of Michigan Biological Station
Professor, Dept. of Ecology & Evolutionary Biology
Department of Ecology and Evolutionary Biology
University of Michigan
830 N. University, 1029 Kraus Nat. Sci. Bldg.
Ann Arbor, MI 48109-1048
Tel: 734-763-4461
Fax: 734-647-1952
Email: knute@umich.edu

Satish Nandapurkar
CEO
Chicago Climate Exchange
190 S. LaSalle St., Ste. 1100
Chicago, IL 60603
Tel: 312-229-5177
Fax: 312-554-3373
Email: satish@chicagoclimateexchange.com

Frank Niepold
Climate Science Literacy Coordinator
UCAR/NOAA Climate Program Office
1315 East-West Hwy., SSMC3
Rm. 12727 North
Silver Spring, MD 20910
Tel: 301-734-1244; Cell: 240-429-0038
Fax: 301-713-0518
Email: frank.niepold@noaa.gov

Dev Niyogi
Associate Professor of Regional Climatology
and Indiana State Climatologist
Department of Agronomy
Department of Earth and Atmospheric Sciences
Purdue University
West Lafayette, IN 47907
Tel: 765-494-6574
Fax:
Email: climate@purdue.edu

Rolf N. Nordstrom
Executive Director
Great Plains Institute
2801 21st Ave., Ste. 220
Minneapolis, MN 55407
Tel: 612-278-7150; Cell: 651-246-9386
Fax:
Email: rnordstrom@gpisd.net

Adam Parris
Program Manager, Regional Integrated Sciences and
Assessments
NOAA Climate Program Office
1315 East-West Highway, SSMC-3
Silver Spring, MD 20910
Tel: 301-734-1243
Fax: 301-713-0518
Email: adam.parris@noaa.gov

Edward A. Parson
Joseph L. Sax Collegiate Professor of Law
Professor of Natural Resources & Environment
University of Michigan
432 Hutchins Hall
625 South State Street
Ann Arbor, MI 48109-1215
Tel: 734-763-6133
Fax: 734-763-9375
Email: parson@umich.edu

Jonathan Patz
Professor
Center for Sustainability and the Global Environ-
ment (SAGE)
University of Wisconsin, Madison
1710 University Ave, Rm. 258
Madison, WI 53726
Tel: 608-262-4775
Fax:
Email: patz@wisc.edu

Craig A. Peterson (Ret.)
Applied Sciences and Technology Project Office
NASA John C. Stennis Space Center
NASA/PA30 Building 1100, Suite 2005H
Stennis Space Center, MS 39529
Tel: 228-688-1984
Fax: 228-688-1399
Email: craig.a.peterson@nasa.gov

Kelly T. Redmond
Regional Climatologist/Deputy Director
Western Regional Climate Center
Desert Research Institute
2215 Raggio Pkwy.
Reno, NV 89512-1095
Tel: 775-674-7011
Fax: 775-674-7016
Email: krwrcc@dri.edu

Ed Roggenkamp
Environmental Law and Policy Center
35 East Wacker Dr., Ste. 1300
Chicago, IL 60601
Tel: 312-673-6500
Fax:
Email: eroggenkamp@elpc.org

Paul J. Runci
Joint Global Change Research Institute
Pacific Northwest National Laboratory
University of Maryland
5825 University Research Ct., Ste. 3500
Baltimore, MD 20740
Tel: 301-314-7843
Fax: 301-314-6719
Email: paul.runci@pnl.gov

Mark Russo
Meteorologist
Chesapeake Energy Corporation
100 North Riverside Plaza, Suite 2350
Chicago, IL 60606
Tel: 312-756-1805
Fax: 312-756-1807
Email: mark.russo@chk.com

Don Scavia
Graham Family Professor and Director
Graham Environmental Sustainability Institute
Professor, School of Nat. Res. & Environment
Professor, Civil and Env. Engineering
University of Michigan
Ann Arbor, MI 48109-1041
Tel: 734-615-4860
Fax:
Email: scavia@umich.edu

Doug Scott
Director
State of Illinois
Environmental Protection Agency
4302 North Main St.
Rockford, IL 61103
Tel: 815-987-7554; Cell: 217-494-2947
Fax: 815-987-7005
Email: doug.scott@illinois.gov

Eileen Shea
Chief, Climate Services Division
National Climatic Data Center, NOAA/NESDIS
Veach-Baley Federal Building, Room 468
151 Patton Avenue
Asheville, NC 28801-5001
Tel: 828-271-4384
Fax: 828-271-4876
Email: eileen.shea@noaa.gov

A.J. Singletary
Office of Safety, Energy and Environment
U.S. Department of Transportation
1200 New Jersey Avenue, SE
Washington, DC 20590
Tel: 202-366-0360
Fax: 202-366-0263
Email: arthur.singletary@dot.gov

Melissa Soline
Program Manager
Great Lakes and St. Lawrence Cities Initiative
177 N. State St., Ste. 500
Chicago, IL 60601
Tel: 312-201-4517
Fax: 312-553-4355
Email: Melissa.soline@glslcities.org

Brooke Stewart
Department of Atmospheric Sciences
University of Illinois
105 S. Gregory Street
Urbana, IL 61801
Tel: 217-333-9604
Fax:
Email: stewar14@atmos.uiuc.edu

Heather M. Stirratt
Great Lakes Regional Coordinator
NOAA Coastal Services Center
1735 Lake Drive West, Building NOHRSC
Chanhassen , MN 55317
Tel: 952-368-2505
Fax:
Email: heather.stirratt@noaa.gov

Deborah Stone
Deputy Director
Illinois Department of Natural Resources
One Natural Resources Way
Springfield, IL 62702
Tel: 217-785-1807
Fax: 217-299-8240
Email: Deborah.stone@illinois.gov

Eugene Takle
Professor
Dept. of Geologicial and Atmospheric Science
Iowa State University
3013 Agronomy Hall
Ames, IA 50011
Tel: 515-294-9871
Fax:
Email: gstakle@iastate.edu

Kate Tomford
Director of Sustainability
Office of Governor Pat Quinn
100 W. Randolph St.
Chicago, IL 60601
Tel: 312-814-4083
Fax: 312-814-4862
Email: kate.tomford@illinois.gov

Bradley H. Udall
Director
Western Water Assessment
University of Colorado at Boulder
NOAA Earth System Research Laboratory
325 Broadway R/PSD
Boulder, CO 80305-3328
Tel: 303-497-4573
Fax: 303-497-6449
Email: Bradley.udall@colorado.edu

Dan Walker
Chief, Climate Assessments and Services Division
NOAA Climate Program Office
1315 East-West Highway, SSMC-3
Silver Spring, MD 20910
Tel: 301-734-1212
Fax: 301-713-0518
Email: daniel.walker@noaa.gov

Margaret Walsh
Climate Change Program Office
U.S. Department of Agriculture
1400 Independence Ave., SW
Washington, DC 20250
Tel: 202-720-9978
Fax: 202-401-1176
Email: mwalsh@oce.usda.gov

Michael Walsh
Executive Vice President
Chicago Climate Exchange
190 S. LaSalle St., Ste. 1100
Chicago, IL 60603
Tel: 312-229-5177
Fax: 312-554-3373
Email: mwalsh@chicagoclimateexchange.com

Charles L. Walthall
National Program Leader, Climate Change, Soils and
Emissions Research Program
Office of National Programs
USDA Agricultural Research Service
5601 Sunnyside Ave., Room 4-2282
Beltsville, MD 20705-5140
Tel: 301-504-4634; Cell: 443-968-0660
Fax: 301-504-6231
Email: charlie.walthall@ars.usda.gov

Anne M. Waple
NOAA National Climate Data Center
Veach-Baley Federal Building
151 Patton Ave.
Asheville, NC 28801
Tel: 828-257-3000
Fax:
Email: anne.waple@noaa.gov

Thomas Wilbanks
ORNL Corporate Fellow
Group Leader, MultiScale Energy-Environmental
Systems, Environmental Sciences Division
Oak Ridge National Laboratory
Bethel Valley Road, Building 1505, Room 356A
Oak Ridge, TN 37831-6038
Tel: 865-574-5515
Fax: 865-576-2943
Email: wilbankstj@ornl.gov

Pammela J. Wright
Board Member
National Farmers Union - Missouri
President, Central Ozark Farmers Union Chapter
9492 County Road 9190
West Plains, MO 65775
Tel: 417-257-1770
Fax:
Email: pammela@socket.net

Donald J. Wuebbles
The Harry E. Preble Professor of Atmospheric Sciences
School of Earth, Society, and Environment
Department of Atmospheric Sciences
University of Illinois
105 S. Gregory St
Urbana, IL 61801-3070
Tel: 217-244-1568
Fax: 217-244-4393
Email: wuebbles@uiuc.edu

John A. Young
Professor Emeritus & Director, Wisconsin State Climatology Office
University of Wisconsin-Madison
1225 W. Dayton St.
1503 Atmospheric and Oceanic Sciences
Madison, WI 53706
Tel: 608-263-2374; Cell: 608-669-0949
Fax: 608-262-0166
Email: jayoung@wisc.edu

James A. Zandlo
State Climatology Office, DNR Waters
439 Borlaug Hall, 1991 Upper Buford Circle
St. Paul, MN 55108-6028
Tel: 651-297-1314
Fax: 651-625-2208
Email: jzandlo@umn.edu

Notes:

www.ingramcontent.com/pod-product-compliance
Lightning Source LLC
Chambersburg PA
CBHW081412170526
45166CB00010B/3312